Bicycle Lanes Versus Wide Curb Lanes:

Operational and Safety Findings and

Countermeasure Recommendations

PUBLICATION NO. FHWA-RD-99-035   OCTOBER 1999

U.S. Department of Transportation
Federal Highway Administration

Research, Development, and Technology
Turner-Fairbank Highway Research Center
6300 Georgetown Pike
McLean, VA 22101-2296

1. Report No. FHWA-RD-99-035	2. Government Accession No.	3. Recipient's Catalog No.
4. Title and Subtitle BICYCLE LANES VERSUS WIDE CURB LANES: OPERATIONAL AND SAFETY FINDINGS AND COUNTERMEASURE RECOMMENDATIONS		5. Report Date
		6. Performing Organization Code
7. Author(s) William W. Hunter, J. Richard Stewart, Jane C. Stutts, Herman H. Huang, and Wayne E. Pein		8. Performing Organization Report No.
9. Performing Organization Name and Address University of North Carolina Highway Safety Research Center 730 Airport Road, CB #3430 Chapel Hill, NC 27599		10. Work Unit No. (TRAIS)
		11. Contract or Grant No. DTFH61-92-C-00138
12. Sponsoring Agency Name and Address Office of Safety and Traffic Operations Research & Development Federal Highway Administration 6300 Georgetown Pike McLean, VA 22101-2296		13. Type of Report and Period Covered Final Report March 1995 - May 1998
		14. Sponsoring Agency Code

15. Supplementary Notes

Contracting Officer's Technical Representative (COTR): Carol Tan Esse, HSR-20

16. Abstract

This report presents operational and safety findings and countermeasure recommendations from a comparative analysis of bicycle lanes versus wide curb lanes. The primary analysis was based on videotapes of almost 4,600 bicyclists in Santa Barbara, CA, Gainesville, FL, and Austin, TX. The videotapes were coded to evaluate operational characteristics and conflicts with motorists, other bicyclists, or pedestrians.

Significant differences in operational behavior and conflicts were found between bike lanes and wide curb lanes but varied depending on the behavior being analyzed. Wrong-way riding and sidewalk riding were much more prevalent at WCL sites compared with BL sites. Significantly more motor vehicles passing bicycles on the left encroached into the adjacent traffic lane from WCL situations compared with BL situations. Proportionally more bicyclists obeyed stop signs at BL sites; however, when a stop sign was disobeyed, the proportion of bicyclists with both "somewhat unsafe " and "definitely unsafe" movements was higher at BL sites. The vast majority of observed bicycle-motor vehicle conflicts were minor, and there were no differences in the conflict severity by type of bicycle facility. Bicyclists in WCLs experienced more bike/pedestrian conflicts while bicyclists in BLs experienced more bike/bike conflicts. An initial model fitted to the intersection conflicts showed no differences in the conflict rate by type of bicycle facility, but showed higher conflict rates for left turn movements.

The overall conclusion is that *both* BL and WCL facilities can and should be used to improve riding conditions for bicyclists. The identified differences in operations and conflicts appeared to be related to the specific destination patterns of bicyclists riding through the intersection areas studied and not to the characteristics of the bicycle facilities.

In addition to this implementation manual, there is a final report (FHWA-RD-99-034) containing a complete discussion of the research method, data collection procedures, and data analysis, as well as a guidebook (FHWA-RD-99-036) about innovative bicycle accommodations.

17. Key Words: Bicycle lane, wide curb lane, bicycle operations, bicycle maneuvers, conflicts	18. Distribution Statement No restrictions. This document is available to the public through the National Technical Information Service, Springfield, Virginia 22161		
19. Security Classif. (of this report) Unclassified	20. Security Classif. (of this page) Unclassified	21. No. of Pages 31	22. Price

Form DOT F 1700.7 (8-72)   **Reproduction of form and completed page is authorized**

Table of Contents

List of Figures

Chapter

1

Introduction

Background

A number of recent events renders a study of bicycle facilities as appropriate and timely. The passage of the 1991 Intermodal Surface Transportation Efficiency Act (ISTEA) legislation meant a variety of funds could be more readily used by local and state officials to plan and build such facilities. Indications are that many governments and agencies have taken advantage of the opportunity. Publication of the National Bicycling and Walking Study in 1994 with the U.S. Department of Transportation (USDOT) goals of doubling the percentage of trips made by bicycling and walking and simultaneously reducing by 10 percent the number of bicyclists and pedestrians injured or killed in traffic crashes adds emphasis to the need to accommodate non-motorists with well-designed facilities. User survey respondents have clearly stated that more facilities are desired and will increase the amount of travel by bicycle.

This research summary is intended to convey operational and safety information developed in a recent study of bicycle lanes (BLs) and wide curb lanes (WCLs). The document is based on the parent study titled *A Comparative Analysis of Bicycle Lanes Versus Wide Curb Lanes* (Hunter, Stewart, Stutts, Huang, and Pein, 1998).

A Brief Discussion of Bicycle Lanes and Wide Curb Lanes

A long-standing issue in the bicycling community centers on whether bicycle lanes or wide curb lanes are preferable. A bicycle lane (BL) is a portion of a roadway that has been designated by striping and pavement

Figure 1. Typical bicycle lane.

markings for the preferential or exclusive use of bicyclists (figure 1).

BL width is normally in the range of 1.2 to 1.8 m. A wide curb lane (WCL) is the lane nearest the curb that is wider than a standard

Figure 2. Typical wide curb lane.

lane and provides extra space so that the lane may be shared by motor vehicles and bicycles (figure 2). Thus, WCLs may be present on normal two-lane roadways or on multilane roadways. A desirable width for WCLs is 4.3 m. Lanes wider than 4.6 m sometimes result in the operation of two motor vehicles side by side. Many bicyclists report feeling safer when riding on BLs, while BL opponents venture that these facilities make it difficult for bicyclists to handle turning maneuvers at intersections, especially left turns. WCL advocates feel that these wider lanes encourage cyclists to operate more like motor vehicles and thus lead to more correct maneuvering at intersections.

Because a WCL is a wider-than-normal traffic lane that is shared with motor vehicles, some do not refer to this layout as a bicycle facility. However, for the purposes of this report, both BLs and WCLs will be referred to as bicycle facilities.

Parent Study Method

Overview

In the parent study, bicyclists in either a BL or WCL were videotaped as they approached and proceeded through eight BL and eight WCL intersections with varying speed and traffic conditions in three cities. Approximately 4,600 bicyclists were videotaped (2,700 riding in BLs and 1,900 in WCLs). The videotapes were coded to learn about operational characteristics (e.g., intersection approach position and subsequent maneuvers) and conflicts with motor vehicles, other bicycles, or pedestrians. A conflict was defined as an interaction between a bicycle and motor vehicle, pedestrian, or other bicycle such that at least one of the parties had to change speed or direction to avoid the other. Both bicyclist and motorist maneuvers in conflict situations were coded and analyzed. This would cover maneuvers such as a bicyclist moving incorrectly from the bicycle lane into the traffic lane prior to making a left turn, or conversely, a motor vehicle passing a bicyclist and then abruptly turning right across its path.

City Selection

Considerable effort in the early part of the project was spent in identifying possible cities for study. Candidates were narrowed and visits made to Santa Barbara, CA; the

Palo Alto area of CA; Madison, WI; Gainesville, FL; and Austin, TX. Based on key factors such as amount and type of facilities, number of riders, willingness and eagerness of local contacts to participate, and windows of opportunity (i.e., climate) for videotaping, Santa Barbara, CA, Gainesville,

Figure 3. City map.

FL, and Austin, TX, were selected as primary project cities (figure 3). These were spread geographically across the United States and provided for a good comparative analysis.

Site Characteristics

The objective was to achieve a group of sites within the cities that varied by width of BL or WCL (two levels), speed limit (two levels), and traffic volume (two levels). Such a matrix yields a total of eight sites. Thus, eight BL and eight WCL sites were selected for videotaping in each city. Selected breakpoint values were:

BL width - 1.5 m or less, >1.5 m
WCL width - 4.3 m or less, >4.3 m
Speed limit - 50 km/h or less, >50 km/h
Traffic volume - Low volume up to 7,500 vpd for 2 lanes; 15,000 vpd for 4 lanes, +etc.

High volume greater than 7,500 vpd
for 2 lanes; 15,000 vpd for 4 lanes, etc.

We also tried to satisfy an objective of having 20 to 30 bicyclists per hour riding through the selected intersections. The BL and WCL matrices shown below provide the overall mix for all three cities combined.

As potential sites were selected in each city, we attempted to develop a mix based on the variable parameters shown above, as well as attempted to have variety in the sites that is representative of real-world conditions (e.g., BL and WCL sites with and without parking, BL sites with a weaving area and a bike pocket, BL sites with and without the stripe carried all the way to the intersection, BL and WCL sites where turning lanes were added at the intersection). In all three cities, however, the preliminary site list of top candidates had to be altered, usually due to a small number of riders available for videotaping. BL sites were generally popular and tended to have a reasonably high number of bicyclists available. Sometimes the preliminary list of BL sites was altered because it was discovered that the viewing angle for videotaping was not good. It was difficult to find eight suitable WCL sites in any of the selected cities due to small numbers of bicyclists riding on WCL facilities.

Bike Lane Sites

Width of BL	1.5 m or less		>1.5 m	
Traffic Volume	Low	High	Low	High
50 km/h or less	FL FL FL CA CA TX TX TX TX TX	FL FL CA	FL FL FL CA TX TX	CA CA
>50 km/h		CA TX		CA

Wide Curb Lane Sites

Width of WCL	4.3 m or less		>4.3 m	
Traffic Volume	Low	High	Low	High
50 km/h or less	FL TX	TX	FL FL TX TX TX	FL FL FL CA CA CA TX TX
>50 km/h		CA TX	CA	FL FL CA CA CA

In Gainesville and Austin, the selected sites were quite close to the university campuses, because this is where the majority of the bicyclists were located, and data could be collected in an efficient manner. In Santa Barbara, the university campus was remote, and student bicyclists were a much smaller part of the mix. In the three cities selected, BL sites tended to concentrate at low traffic speed and low traffic volume locations, while WCL sites tended to concentrate at high traffic volume locations. Overall, the matrices of final sites indicate a reasonable mix of variation.

Besides the items mentioned above, a variety of other descriptive data items were collected at each site. These included type of area, pavement marking (striping) information for the BLs and WCLs, traffic control device present, number of lanes, estimated driving speed, presence of parking, average annual daily traffic (AADT), and others.

Videotaping of Bicyclists

The initial plan was to videotape bicyclists both at midblock and intersection locations. However, it became apparent that sample sizes would be relatively small if the videotaping task was divided in this fashion. Thus, the decision was made to forego the midblock videotaping and instead videotape each intersection twice for the following reasons:

- Intersections account for about half of all bicycle-motor vehicle crashes.
- Because of the need to make turning movements, intersections were expected to lead to more conflicts between vehicles, pedestrians, and other bicycles.
- It was of interest to learn about the maneuvers bicyclists make to travel through intersections, such as the ways left turns are made.
- The camera position would allow viewing of the approaching bicyclists from a

considerable distance back from the intersection (not unlike a midblock situation).

Intersections and the approaches to intersections (referred to as midblock hereafter) were thus the focus of the data collection effort. Bicyclists were videotaped in the oncoming direction as they approached the selected intersections. The two-person data collection team usually mounted the camera on a 3-m stepladder set up some 30 to 40 m on the far side of the intersection. The location was such that the oncoming bicyclists generally were not aware of the camera until close to the intersection. The stepladder was quite beneficial in providing a viewing angle above traffic. In a few of the Gainesville intersections, a platform truck belonging to the city was used to enable a better viewing position than could be afforded by a stepladder.

Normally the camera position allowed for a view of more than 150 m back from the intersection. Five 46-cm traffic cones were set up at 30-m intervals from the intersection stop bar location (at 30, 60, 90, 120, and 150 m). Approaching bicyclists were usually captured before reaching the 150 m cone and followed through the intersection (figure 4). The data collector would zoom in on the bicyclist to enable a better view of any kinds of bicycle-motor vehicle interactions. If the bicyclist had to stop for a traffic signal, the data collector would ascertain if it were possible to videotape another approaching bicyclist. If so, this bicyclist would be followed up to the intersection, and then both bicyclists would be taped as they rode through. Each intersection was videotaped twice for two hours at each session.

Generally all 16 sites in a city were videotaped once before the second round

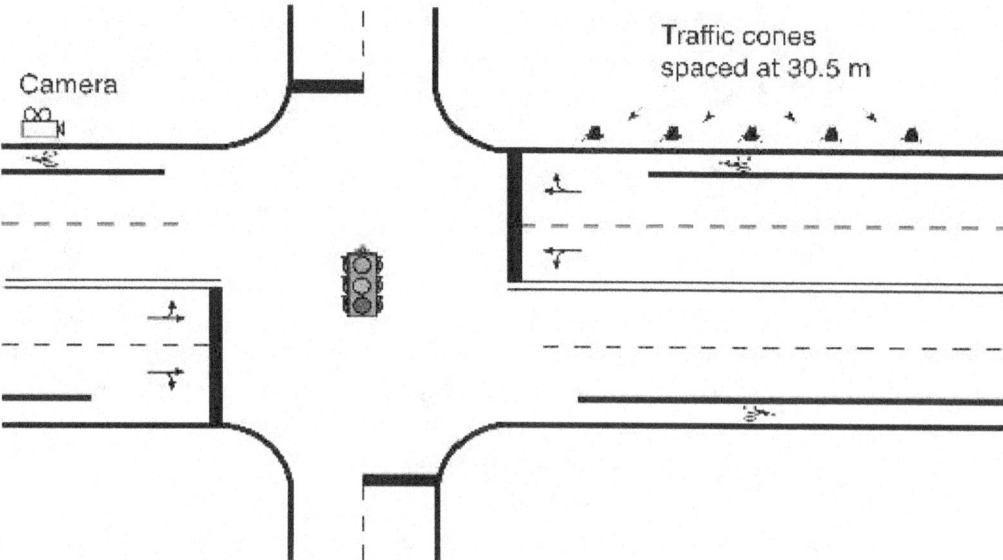

Camera

Traffic cones
spaced at 30.5 m

Figure 4. Typical data collection set-up.

of taping began. As stated earlier, approximately 4,600 bicyclists were videotaped in the three cities (2,700 at BL sites and 1,900 at WCL sites).

traffic were also videotaped that corresponded to the time of the bicycle videotaping. The camera was positioned at a location where all the legs of the intersection could be observed. This videotape enabled counts of motor vehicles traveling through the intersection and thus some measure of exposure to traffic.

Bicyclist Experience Data

Bicyclist experience data were also collected separately from the videotaping at each of the 16 data collection sites in each city through use of a short oral survey. Four questions were asked:

● What is your age?

● On average, how many days a week do you ride your bike?

● On average, about how many miles do you ride each week?

● How would you classify yourself with respect to the experience you have riding on city streets? (1 or 2, shown following)

Besides the bicyclist videotaping described above, 15-minute samples of

1. I feel comfortable riding under most traffic conditions, including major streets with busy traffic and higher speeds.

OR

2. I only feel comfortable riding on streets with less traffic and lower speeds, on streets with bicycle lanes, or on sidewalks.

In addition, information was coded pertaining to where the bicyclists were riding (road, sidewalk, or other location) as they approached the survey station, and the race, gender, and helmet use of the bicyclists. The data enabled information about the experience level of bicyclists riding through the particular intersection. Such knowledge could be directly relevant to the types of maneuvers and conflicts seen at the site. Each experience data collection session lasted 2 hours and was matched to the videotaping (i.e., same basic time of day and day of week) but was always done a few days before or after the videotaping.

Slightly more than 2,900 surveys were completed. These data were analyzed to learn about the age, riding habits, and experience levels of the bicyclists riding through these intersections.

Coding of Videotape Data

A form for coding a variety of items associated with a bicyclist approaching and riding through an intersection was developed, tested, and revised several times before the form was satisfactory. The objective was to code actions associated both with a "midblock" (the intersection approach) and an intersection area. Midblock was thus defined as the area between the third and fifth traffic cones set up on the approach leg (90-150 m from the intersection stop bar location). The intersection was defined as the area covered by the first three traffic cones (0-90 m back from the stop bar location).

The following are examples of the types of variables that were coded:
- Bicyclist riding wrong way.
- Bicyclist demographics and helmet use.
- Midblock positions and movements.
- Bicyclist spacing from the curb or gutter pan seam and from a passing motor vehicle.
- Bicyclist midblock behaviors (e.g., turning across a lane of traffic).
- Midblock conflict information.
- Intersection positions and movements.
- Bicyclist straight, left turn, and right turn methods.
- Bicyclist straight, left turn, and right turn conflict information.

Coding and Analysis of Crash Data

Two years of recent (1994 and 1995)

bicycle-motor vehicle crash data were obtained from each of the three cities. Crashes from one complete year (1995) from each city were "typed" following the methodology originally developed by the National Highway Traffic Safety

Administration (NHTSA) in the late 1970s[1] and being modified in partnership with the Federal Highway Administration (FHWA) for computer application. The computer software will be known as PBCAT (Pedestrian and Bicycle Crash Analysis Tool), a user-friendly software package developed for FHWA by the University of North Carolina Highway Safety Research Center. Crashes from each city were relatively sparse, and very few matched the intersections selected for videotaping. However, city trends could be examined to determine if overall crash patterns were similar to the types of behaviors and conflicts coded from the videotape data.

Organization of the Manual

Chapter 2 provides summary information from the parent study from all of the data sources listed above, including videotaping, bicyclist experience survey, and crash data. Problem situations found in BLs and WCLs are then highlighted. Chapter 3 recommends countermeasures for the problem situations, and Chapter 4 offers concluding comments.

---

[1]For more specific background on crash typing, see Hunter, Stutts, Pein, and Cox (1996).

Chapter 2
Summary of Main Results

This comparative analysis of BL and WCL sites was based on videotapes of almost 4,600 bicyclists in three U.S. cities approaching and then riding through intersections for which the associated bicycle facility was either a BL or WCL. In two of the three cities, the vast majority of bicyclists were traveling to or from college campuses, and the intersections selected were generally in bicycle commuting corridors. The intent was to videotape bicyclists who regularly ride in traffic. The result was a group of sites with varying "real-world" characteristics such as different BL striping techniques (e.g., using a solid or dashed BL stripe all the way to the intersection), presence of parking (e.g., a combination BL and parking lane), and provision of turn lanes at intersections that sometimes narrow the nominal width of the BL or WCL at the intersection proper. What follows is a brief summary of the main operational and safety (conflict) results and some further elaboration of key issues.

Summary of Main Results

Bicyclist Characteristics

• The overwhelming majority of videotaped bicyclists were between the ages of 16 and 64. Slightly more than three-fourths were male.

• While wrong-way riding on a sidewalk is not necessarily illegal or improper behavior, it can lead to operational and safety problems with motor vehicle traffic. Thus, it has been defined and used in this report as a behavioral characteristic of bicyclists. Overall, 5.6 percent of the bicyclists were riding the wrong way (i.e., facing traffic). This included 1.3 percent in the road and 4.3 percent on sidewalks. However, wrong-way riding was much more prevalent on the sidewalk at WCL sites (7.0 percent) compared with BL sites (2.3 percent). Eliminating sidewalk riding from the comparison, however, still resulted in significantly more wrong-way riding associated with WCL sites (1.7 percent) than BL sites (1.0 percent).

• A bicyclist experience oral survey was administered to bicyclists proceeding through the project sites on days when videotaping was not being done. There were no statistically significant differences in the age, gender, and helmet use of bicyclists by type of facility. Higher proportions of Whites and Blacks rode in WCL situations and higher proportions of Asians and Hispanics in BL situations, and the differences were significant.

• Bicyclists surveyed at WCL sites tended to ride more days per week, but the miles per week for bicyclists at BL versus WCL sites were equivalent. Overall about one-third of the riders at both BL and WCL sites considered themselves to be experienced bicyclists.

• When bicyclists were surveyed, their riding location (i.e., in the street or on the sidewalk) when approaching the survey station was recorded. Surveyed bicyclists showed the same tendency as the videotaped bicyclists in that sidewalk riding was more associated with WCL sites.

Midblock Movements

• In the midblock or intersection approach area (between 90 and 150 m from the intersection), significantly more motor vehicles passing bicycles on the left encroached into the adjacent motor vehicle traffic lane from WCL situations (17 percent) compared with BL situations (7 percent). This is in agreement with results

Figure 5. Vehicle encroachment into the adjacent travel lane when passing bicyclists in a WCL.

from a recent Florida DOT study (Harkey and Stewart, 1997). However, encroachments into the adjacent traffic lane very

rarely resulted in a conflict with another motor vehicle (figure 5).

Statistical Modeling of Spacing Between Bicycles and Motor Vehicles

Using least squares regression analysis, statistical models were used to examine lateral positioning of the bicyclists with respect to the curb and parked vehicles, as well as separation distance between bicyclists and motor vehicles. The primary purpose of this analysis was to determine which geometric and traffic operational variables influence these measures and to determine if there were differences in these measures that could be attributed to type of facility (i.e., wide curb lane vs. bicycle lane). The results from this analysis are summarized below.

• On facilities with no on-street parking, bicyclists tended to position themselves closer to the curb (or gutter pan seam, if present) when the BL widths were less than

or equal to 1.6 m compared with WCL facilities with the same traffic volume. When the BLs were greater than 1.6 m in width, bicyclists tended to position themselves further from the curb compared with WCL sites. When motor vehicles were passing bicyclists, the position of the bicyclists tended to be about 0.3 m closer to the curb compared with their position when not being passed. This result was the same irrespective of type of facility.

• On roadways with bicycle lanes and on-street parking, bicyclists positioned themselves about the same distance from parked vehicles as they did from the curb on roadways with bicycle lanes and no on-street parking. The small number of observations on WCL facilities with on-street parking prohibited similar analyses for WCL sites.

• With respect to separation distance between bicyclists and passing motor vehicles, there were no practical differences between BL sites and WCL sites. Instead, this distance was primarily a function of the total width available (either the WCL width or the BL width and adjacent motor vehicle lane width combined). (See Harkey and Stewart (1997) for more information about spacing between bicycles and motor vehicles at midblock locations when the bicycle facility is either a BL, WCL, or paved shoulder.)

Intersection Movements

• The intersection was defined as starting 90 m upstream from the stop bar and included the intersection proper. Proportionally more bicyclists approached the intersection on a sidewalk when the facility was a WCL (15 percent) than a BL (3 percent).

• Overall, 92 percent of bicyclists obeyed the traffic signals that were present, and there were no differences by facility type. When a signal was disobeyed, 16 percent of the actions were considered somewhat unsafe and 2 percent definitely unsafe. There were no differences by facility type.

• Overall, 75 percent of bicyclists obeyed existing stop signs. Proportionally more bicyclists obeyed stop signs at BL sites (81 percent) than at WCL sites (55 percent). When a stop sign was disobeyed, 13 percent were considered somewhat unsafe and 2 percent definitely unsafe. The proportion of bicyclists with both somewhat unsafe (19 versus 5 percent) and definitely unsafe (3 versus 0 percent) movements was higher at BL sites. The differences between BL and WCL sites were significant when the somewhat unsafe and definitely unsafe categories were combined.

• Seventy-two percent of the bicyclists went straight through the intersection, with another 15 percent turning left and 13 percent turning right. There were no differences by facility type. Nine percent of the bicyclists tended to shy to the right (i.e., move to the right and away from traffic) as they went straight through the intersection (11 percent in BLs and 7 percent in WCLs), and this difference was significant.

• Left turns presented a problem for bicyclists and were made in a variety of ways (figure 6). Overall, 44 percent made left turns

Figure 6. Bicyclist making a left turn in advance of the intersection.

like a motor vehicle with *proper* lane destination positioning (41 percent from BL sites and 48 percent from WCL sites). On the other hand, 14 percent of bicyclists at WCL sites made motor vehicle style left turns with *improper* lane destination positioning compared with 3 percent from BL sites. There were proportionally more pedestrian style left turns from WCL sites (24 percent versus 12 percent from BL sites). Both findings may reflect the generally higher traffic volumes and speeds and greater number of lanes at WCL sites.

• Right turns for bicyclists were an easier maneuver, with only 13 percent made in a non-standard fashion (e.g., from a BL or WCL to a wrong way position on the cross street). Nineteen percent of the right turns made at WCL sites were non-standard versus 10 percent of right turns at BL sites, and the differences were significant.

Midblock Conflicts

A conflict was defined as an interaction between a bicycle and motor vehicle, pedestrian, or other bicycle such that at least one of the parties had to change speed or direction to avoid the other.

• Of the 188 midblock conflicts, 71 percent were bicycle/motor vehicle, 10 percent bicycle/bicycle, and 19 percent bicycle/pedestrian. Almost all of the bike/bike conflicts occurred in BLs, typically due to one bicyclist maneuvering around a slower moving bicyclist. Compared with BLs, bicyclists in WCLs experienced more bike/pedestrian conflicts (30 percent versus 16 percent, and reflective of the increased sidewalk riding in WCL situations) and less bike/bike conflicts. The differences by facility type were statistically significant.

• There were no differences in the bicycle or motor vehicle avoidance response scales by facility type. The scales ranged from no change in riding or driving up to

Figure 7. Bicyclist swerving across a lane of traffic.

collision or near crash.

● Overall, 98 percent of the midblock conflicts were coded as minor, and there were no differences by facility type.

● Bicycle actions more associated with BLs in these midblock conflicts included the bicycle having to slow, stop, or swerve for traffic not influenced by the intersection; the bicycle turning or swerving across a lane of traffic (figure 7); encounters with other bikes; and "other" bike actions (such as an improper left turn). The bicycle action more associated with WCLs in these midblock conflicts was encounters with pedestrians.

● Motor vehicle actions more associated with BLs in these midblock conflicts included illegal parking in the BL and entering/exiting on-street parking or a driver or passenger entering/exiting a parked or stopped vehicle. Motor vehicle actions more associated with WCL conflicts included turning right in front of a bicyclist after overtaking and "other" actions such as failing to yield, improper right turns, and crowding bikes.

Intersection Conflicts

Similar to the midblock area, an intersection conflict was defined as an interaction between a bicycle and motor vehicle, pedestrian, or other bicycle such that at least one of the parties had to change speed or direction to avoid the other.

● Of the 198 intersection conflicts, 79 percent were bike/motor vehicle, 10 percent bike/bike, and 10 percent bike/pedestrian. The differences in the BL/WCL distributions were statistically significant. There were proportionally more bike/bike conflicts in BLs (15 percent) and less in WCLs (4 percent). Conversely, there were proportionally more bike/pedestrian conflicts in WCLs (17 percent, and again reflective of sidewalk riding) and less in BLs (6 percent).

● The position of the motor vehicle with respect to the bicycle in the intersection conflicts was 66 percent in the same direction, 6 percent in the opposing direction, 5 percent approaching from the left, 15 percent approaching from the right, and 7 percent approaching from some other position. There were no differences by facility type.

● There were no differences in the bicycle or motor vehicle avoidance response scales by facility type.

● Overall, 93 percent of the intersection conflicts were coded as minor, and there were no differences by facility type.

● Bicycle actions more associated with BLs in these intersection conflicts included the bicycle having to slow/stop/swerve for intersection traffic, the bicycle having to slow/stop/swerve for traffic not influenced by the intersection, and the bicycle turning or swerving across a lane of traffic. Bicycle actions more associated with WCLs included passing slow moving or stopped vehicles on the right, encounters with pedestrians, and "other" actions such as improper left turns and merging onto the road from a sidewalk.

● Motor vehicle actions more associated with BLs included illegal parking in the BL and "other" actions such as a driver or passenger entering/exiting a parked or stopped vehicle (figure 8) and crowding the BL. Motor vehicle actions more associated with WCLs included having to slow/stop/swerve for intersection traffic and turning right in front of a bicyclist after overtaking.

Statistical Modeling of Conflict Data

● Raw frequency bike-motor vehicle conflict rates per entering bicyclist were slightly higher at BL sites than WCL sites when midblock and intersection conflict data were combined (6.7 versus 5.1 bike-motor vehicle conflicts per 100 entering bicyclists).

● The rate of *midblock* bike/motor vehicle conflicts associated with BLs was considerably higher than the rate for WCLs, although the rates were small. Generalized linear models fitted to the data showed that both the presence of a BL and the BL width, along with traffic volume and the presence of driveways, were significant variables in the midblock conflict rate models. The practical effect of such models was that the midblock bike/motor vehicle conflict rate was higher at sites with BLs less than 2.5 m wide than at WCL sites. However, a closer examination of the data revealed that the higher midblock BL conflict rates were attributable to only a few sites. The midblock conflicts at the 10 highest rate sites were thus examined clinically.

● An initial model fitted to the *intersection* conflicts showed no differences in the conflict rate by type of bicycle facility, but higher conflict rates for bicycle left turn movements. A subsequent model was developed that included different intersection types (figure 9) based on the type of BL striping (e.g., solid stripe to the intersection, dashed stripe to the intersection) and whether the typical WCL cross section was maintained through the intersection (or narrowed due to the provision of turn lanes).

The model showed *lower* conflict rates for straight through and right turning bicycles where the BL stripe continued all the way to the intersection and the WCL was not narrowed at the intersection. This is perhaps not surprising, in that bicycles would have more space in these configurations.

As before, a closer study of the data showed that the findings from this model were mainly attributable to a few sites. The difficulty of statistically interpreting outcomes that seemed so dependent on site-specific characteristics led to clinical analysis of higher conflict rate sites, both at midblock and intersection locations. The results of this clinical examination are described below.

Clinical Examination of High Conflict Rate Sites

● The 10 highest conflict rate sites for both the midblock and intersection areas were examined clinically to determine if any typical conflict patterns existed. In the *midblock* area, there were seven BL and three WCL sites. The predominant motor vehicle actions in the midblock conflicts pertained to motor vehicles entering or exiting on-street parking (there were several sites where parking was part of the facility), parking or stopping in the bicycle facilities to let a

Figure 8. Bicyclist having to swerve to avoid the opening of a motor vehicle door.

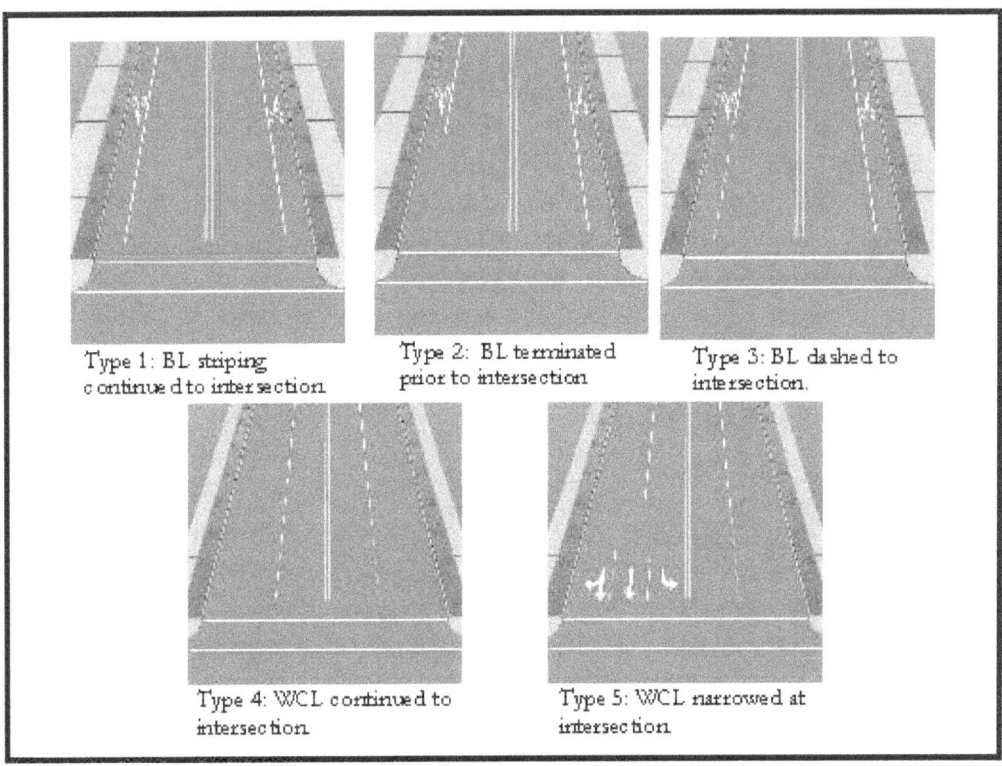

Figure 9. BL and WCL intersection types.

passenger enter or exit the vehicle, and pulling across the BL or WCL into an intersecting street or driveway. The predominant bicycle actions were turning or swerving across a lane of traffic (usually to avoid making a left turn at the intersection ahead) and interacting with pedestrians when riding on the sidewalk. If "fault" in the conflicts had been assigned, the large majority of the fault would have been due to motor vehicle actions.

• In the *intersection* area, there were four BL and six WCL sites. The predominant motor vehicle actions again pertained to entering or exiting on-street parking and parking or stopping in the bicycle facility to let a passenger enter or exit the vehicle. The predominant bicycle actions were turning or swerving across a lane of traffic, passing slow or stopped motor vehicles on the right, and interacting with pedestrians. Some of the conflicts resulted simply from the typical

maneuvering that might occur when bicycles and motor vehicles position themselves to make turns at intersections. If "fault" in the conflicts had been assigned, the majority would have been due to bicycle actions.

• Identifiable situations leading to conflicts from this clinical analysis were presence of parked motor vehicles (either entering/exiting legal parking or illegal parking/stopping) in the BL or WCL, presence of driveways or intersecting streets, and provision of turn lanes at intersections that typically (but not always) resulted in a narrowing of the BL or WCL at the intersection proper (normally in the last 30 to 50 m before the stop bar). Except for combined BL and parking facilities, these situations did not appear to be related to whether a BL or WCL was present. In other words, the conflicts that resulted were site-specific and likely would have occurred whether a BL or WCL was present.

Clinical Examination of Serious Conflicts

● Seventeen conflicts were coded as serious, 10 at WCL and 7 at BL sites. If "fault" had been assigned, 11 would have been the fault of the motorist and 6 the fault of the bicyclist. The motorist turned right soon after overtaking the bicyclist in six of the conflicts (figure 10), pulled from a driveway to the street in three conflicts, and was involved in a parking situation in the other two cases. The bicyclist turned or

Figure 10. Motor vehicle turning right soon after overtaking bicyclists.

swerved across a lane of traffic in three conflicts, disobeyed a traffic signal in two cases, and shifted in front of a motor vehicle in the process of avoiding rough pavement in the other (figure 11). Examining these situations clinically, there appeared to be no differences between BL and WCL serious conflicts.

Comparisons with Crash Data

● One year (1995) of police-reported crash data were "typed" using the NHTSA methodology for all three of the project communities. There were parallels to the videotape data.

●   In Santa Barbara,

one of the two most frequently occurring crash types was the bicyclist striking a parked vehicle. Santa Barbara had a number of individual intersections where parking was part of the bike facility (figure 12), and, overall, 41 percent of the bicyclists were recorded as riding next to parked vehicles, as compared with 21 percent of the Austin bicyclists and none of the Gainesville bicyclists.

● In Gainesville, the most frequently occurring crash type was Motorist Drive Out at Stop Sign. In three out of four of these crashes, the bicyclist was riding the wrong way (facing traffic) on the sidewalk. Seventeen percent of the Gainesville

Figure 11. Bicyclist shifting left to avoid rough pavement.

Figure 12. Bicyclist riding next to parked vehicles in a BL.

bicyclists were observed approaching the targeted intersections on the sidewalk, as compared with less than 3 percent of the Austin bicyclists and less than 2 percent of the Santa Barbara bicyclists. In addition, 9 percent of the Gainesville bicyclists were observed riding the wrong direction on a sidewalk, compared with 1 percent of both the Austin and Santa Barbara bicyclists. Wrong-way sidewalk riding was also a factor in 87 percent of Gainesville's Right Turn on Red crashes, and 75 percent of its Drive Out at Midblock crashes.

● In Austin, 11 percent of the bicyclists made "advance crossovers" to the left prior to the intersection (figure 13), as compared with 3 percent in both Gainesville and Santa Barbara. Nearly 6 percent of the crashes reported for Austin were Bicyclist Left Turn in Front of Motorist. None of these types of crashes were reported for Gainesville or Santa Barbara.

Further Comment

Level of Experience

Many in the bicycling community have assumed that more experienced bicyclists tend to use WCLs and that lesser experienced bicyclists use BLs. This issue was explored in this project by use of an oral questionnaire, where each surveyed bicyclist was asked to read or listen to a statement being read to them about their experience or comfort level on certain types of facilities. Overall results showed that 34 percent of the bicyclists considered themselves to be experienced, and there were no differences by type of facility.

Wrong-Way Riding

Wrong-way riding, or riding facing traffic, was present for approximately 6 percent of the videotaped bicyclists. There seems to be a prevailing feeling that this practice is more widespread in BLs, but in this study a higher proportion of the wrong-way riding tended to occur at WCL sites, whether in the roadway or on the sidewalk (figure 14). Proportionally more of the WCL wrong-way riding took place on the sidewalk; however, eliminating sidewalk riding from the tabulation still showed significantly more wrong-way riding in the street associated with WCL sites. This may be related to the fact that WCLs are often associated with higher volume roadways and that maneuvering through intersections on these roadways can be a complex task. Thus, the bicyclist may choose what seems to be a safer route by riding the wrong way on an adjacent sidewalk or in the street. It may not be safer in actuality, as wrong-way riding either in the street or on a sidewalk is a frequent factor in bicycle-motor vehicle crashes (See Hunter, Stutts, Pein, and Cox, 1996).

PM 5:57
JUN. 18 1997

Figure 13. Bicyclist attempting to make a left turn in front of a motorist.

wide curb lane

Figure 14. A higher proportion of riding facing traffic, both on the sidewalk and in the street, occurred in WCL situations.

Turning and Other Maneuvers at Intersections

Besides the sidewalk riding mentioned above, complexity of traffic at the WCL intersections in this study may also be related to the operational findings that more incorrect left-turn destination positioning and pedestrian-style left turns were associated with WCL intersections. In addition, WCL sites had proportionally more non-standard right turns than BL sites. Left turns presented problems at BL sites as well. An intersection conflict model showed higher conflict rates for straight and right turning bicycles where the bike lane was terminated prior to the intersection, dashed to the intersection, or the nominal width of the BL or WCL was narrowed due to the provision of turn lanes. A prevalent conflict in these situations, whether at a BL or WCL site, is for a motor vehicle to pass a bicyclist and then

turn right soon after the overtaking maneuver is made. Experienced bicyclists can prevent some of these conflicts by taking control of the lane with their positioning, particularly within the intersection, so that the motor vehicle cannot pass. More bicyclists need training related both to turning maneuvers at intersections and to safely negotiating these areas if merely going straight through. Intersections continue to account for about half of all bicycle-motor vehicle crashes (Hunter, Stutts, Pein, and Cox, 1996).

Conflicts

There were nearly 400 midblock and intersection conflicts noted, but the vast majority were minor in nature. There was no difference in the severity level of the conflicts for BL versus WCL sites as measured by bicycle or motor vehicle response scales to conflicts. Bike/bike conflicts were more associated with BLs, while bike/pedestrian conflicts were more associated with WCLs. Unadjusted conflict rates showed BL sites to have slightly higher rates per entering bicyclist than WCL sites.

Many midblock and intersection conflict models were attempted to identify significant variables related to the occurrence of conflicts. A *midblock* conflict model showed that presence and width of a BL were significantly related to conflicts, along with traffic volume and presence of driveways. Conflicts increased with traffic volume, number of driveways, presence of a BL, and narrower BLs. The interpretation question was whether the higher conflict rates were really attributable to these variables, particularly narrower BLs, or to site-specific characteristics for a few locations. Further analysis showed that a few sites with narrower BLs and high conflict rates tended to greatly affect the results. This led to a clinical analysis of high conflict rate sites.

Results of this clinical analysis showed several factors to be consistently related to the occurrence of the conflicts: (1) presence of parked motor vehicles (either entering or exiting legal parking or illegal parking or stopping) in the BL or WCL, (2) presence of driveways or intersecting streets, and (3) provision of additional (usually turn) lanes at intersections that typically (but not always) resulted in a narrowing of the BL or WCL. Fortunately, these are factors for which some countermeasures are available.

Chapter 3

Recommended Countermeasures for Certain Problem Situations Associated with Bike Lanes & Wide Curb Lanes

Problem Situations

The parent study showed several factors to be consistently related to the occurrence of bicycle-motor vehicle conflicts: (1) presence of parked motor vehicles (either entering or exiting a legal parking space or illegal parking or stopping) in the BL or WCL, (2) presence of driveways or intersecting streets, and (3) provision of additional (usually turn) lanes at intersections that typically (but not always) resulted in a narrowing of the BL or WCL. A discussion of recommended countermeasures for these problem situations follows.

Parked Motor Vehicles

Motor vehicle parking conditions vary widely, and there can be large differences between all day parking with low turnover and high turnover parking that typically serves retail stores. High turnover from on-street parking was one of the situations that led to conflicts with bicycles in this study. The other problem situation was illegal parking or stopping in the bicycle facility.

Many communities in the United States allow motor vehicles to park in bicycle facilities (particularly BLs) during some portions of the day, generally when bicycle traffic is low. In other words, there is no bicycle facility when motor vehicles are allowed to park. This practice can only function effectively if police enforcement keeps the motor vehicles out of the facility during the time parking is prohibited. However, this kind of enforcement is difficult to maintain, and violations of these parking provisions are apparent even in bicycle-friendly communities. Eliminating parking altogether in the bicycle facility is a much stronger statement. If bicycling is to be a truly integrated and useful form of transportation, then bicyclists should have facilities available throughout the day.

In like fashion, motor vehicles do not hesitate to pull into BLs to allow passengers to enter or exit. In areas of busy bicycle traffic, this can lead to many conflicts. At the least, standard "no parking in bike lane"

Figure 116: Signs R7-9 and R7-9a

Figure 15. Standard no parking signs for bike lanes.
Source: Oregon Bicycle and Pedestrian Plan, 1995

signs (figure 15) should be used liberally. More often than not, however, this is an enforcement issue.

Besides enforcement, good design policy can help to eliminate some of the conflicts. If motor vehicle parking is an intended part of a BL (i.e., a combination BL with parking), then a double-striped

Figure 16. Double striped BL with parking.

1.5- m BL that positions the right most BL stripe at least 0.9 m from parked vehicles is recommended to provide the best channelization of bicyclists (figure 16). At least 2.4 m should be allowed for parking. When available right-of-way does not allow the double striped BL described above, then a combination lane, intended for both motor vehicle parking and bicycle use, is an alternative. Such a lane should be at least 3.7 m wide, with 4.3 m being preferable, and

Figure 17. Combination BL with parking T's.

contain parking T's (sometimes referred to as tick marks) to denote the parking spaces (figure 17).

Bicyclist education about correct position when riding on streets with on-street parking is also highly recommended. Bicyclists should be at least 0.9 m from parked vehicles, and

riding should be in a straight line. Such recommendations can be easily highlighted on a community bicycle map.

Driveways and Intersecting Streets

Driveways and intersecting streets in either the midblock or intersection area can lead to bike/motor vehicle conflicts. Driveways or alleys in commercial areas are normally the culprits because more motor vehicle traffic is present. Sometimes the problem is the motorist driving out of a driveway or alley and failing to stop before crossing a sidewalk or an implied sidewalk area that has bicycle travel (figure 18). Clear sight lines should be provided for the motorist if possible. If the sidewalk ends at the driveway cut, a crosswalk could be painted (with optional advance stop bar), or the sidewalk could be extended across the driveway cut. A "WATCH FOR BICYCLISTS" sign could also be installed.

Treatments can also be developed for the bicyclist riding on the sidewalk. First and foremost, education should be provided about the dangers of sidewalk riding, and especially wrong-way riding that places the bicyclist out of the normal viewing pattern for a motorist exiting from a commercial driveway or alley. Bicyclists should also be cautioned to ride slowly in these areas that are primarily designed for walking speeds. Painting "USE CAUTION" on the sidewalk at hazardous driveways is also recommended (figure 18).

Most of the problems noted at the high conflict rate sites in this project involved bicyclists riding in the street, however, and not on the sidewalk. From anecdotal observation, it would seem as though motorists are not hesitant to use a BL as a buffer when they exit from a commercial driveway or alley into the street. A remedy is to provide a stop bar for the motorist prior to the BL. "YIELD TO

Figure 18. Possible sidewalk and sign treatments at a busy commercial driveway or alley intersection.

BICYCLISTS" signs may also be helpful (figure 19). Dashing the BL stripe at busy driveways is also recommended, not only to alert a motorist that a bicyclist may be approaching because of the presence of the BL but also to alert a bicyclist that a motorist may be emerging from the driveway adjacent to the dashed stripe.

Equally important is the problem of motorist overtaking where a right turn is made into the driveway soon after the overtaking is completed. The presence of the dashed BL stripe may also help to eliminate some of the serious conflicts and crashes that result from this maneuver. Bicyclist education about the danger of driveways is warranted, with the message focusing not only on motorist-drive-out but also on motorist-overtaking situations. Motorist education relating to the overtaking situation above is also needed.

Additional Lanes at Intersections

There are several problems with additional lanes at intersections. One has to do with the loss of space to the BL or WCL when additional turn lanes are provided with the same width of cross section. It is common practice now to use narrower lanes for turning movements or to calm traffic. Using narrower widths may retain the full width of the bike facility at the intersection.

Another practice involves terminating either the BL or WCL and splitting the approaching traffic into two through lanes just prior to the intersection stop bar area (figure 20). When this occurs, the two lanes often become one again on the far side of the intersection. The idea is to use the extra lane to get traffic through the intersection faster, but along with the notion comes problems for the bicyclist. First is the loss of space. Second is the weaving among the

Figure 19. Signing, marking, and BL dashing treatments at a busy driveway or alley intersection.

motorists as they jockey for position through the intersection and beyond as they must merge again.

Right turn lanes present another problem for bicyclists. There may be weaving between bicycles and motor vehicles in the approach to the right turn in a designated BL if there is a high volume of right turning motor vehicle traffic. Use of a dashed stripe should be beneficial because the stripe gives notice that weaving will take place (figure 21). Bicyclists may also have a tendency to overtake or stop on the right of stopped or slow moving motor vehicles turning right. Education on the hazards associated with this maneuver is recommended.

Whether or not right turn lanes are

Figure 20. Example of traffic splitting.

Figure 21. Dashed BL stripe at right-turn situation.

present, right turning motor vehicles at intersections pose a problem for bicyclists. Similar to the driveway conflict mentioned above, in the intersection area a motor vehicle may also turn right to another street

Figure 22. European bike box.

soon after overtaking a bicyclist. One treatment made popular in Europe that helps to counter this problem is the use of an advanced stop bar or bike box (the term now frequently used to refer to this treatment in the United States). In Europe, a bike box is typically placed at the end of a BL (figure 22) so that bicyclists may proceed easily to the head of the traffic queue and thus get through the intersection ahead of right turning motor vehicle traffic. The bike box is gaining some popularity in the United States, and different versions of the technique are being tried or considered, such as placing a bike box at a WCL intersection, or using the bike box at a BL location to get left turning bicyclists to the head of the queue. The bike box appears to be well accepted in Europe. However, evaluations of such applications in the United States need to be made to determine if the applications are understood and accepted.

Recent Countermeasure Research Findings from Other Countries

There is quite a bit if research available from other countries indicating that the use of symbols, color, and other devices reduce conflicts and crashes at intersections. Almost all of these studies pertain to BLs. It is likely that the use of the markings and color alerts other road users to the presence of bicycle traffic. Brief summaries are provided below:

● In a study in Denmark, Jensen et al., (1997) found that the *marking of bicycle travel paths with blue paint and/or raised pavement* at signalized intersections resulted in 36 percent fewer bicycle-motor vehicle crashes and 57 percent fewer bicyclists who were killed or severely injured.

● At five intersections in Montreal, *colored bicycle crossings* were installed, with the pavement colored blue at bicycle path crossing points. After the markings were painted, bicyclists were more likely to obey stop signs and to stay on designated cycle path crossings. Improved bicyclist behavior led to a decline in the level of conflict between cyclists and motorists (Pronovost and Lusginan, 1996).

● A *raised and painted bicycle path (crossing)* introduced at 44 intersections in Gothenburg, Sweden, reduced motor vehicle speeds (by 35 to 40 percent for right-turning motor vehicles) and increased cyclist speeds (by 10 to 15 percent). The safety improvement was estimated by using a quantitative model and by surveying bicyclists and experts. The model estimated the combined effect of lower motorist speeds and higher bicyclist speeds to be a 10 percent reduction in the number of bicycle-motor vehicle crashes. Bicyclists perceived a 20 percent improvement in safety after the bicycle path was raised and painted. Experts estimated a 30 percent improvement in safety. However, the authors suggested that the total numbers of crashes should be expected to increase due to a 50 percent increase in bicyclists using the improved crossings (Leden, 1997). A follow-on paper using a Bayesian approach for combining the results of the model and surveys estimated a risk reduction of approximately 30 percent attributable to the raised and painted crossing (Gårder, Leden, and Pulkkinen, 1998).

- A variety of other countermeasures, such as traffic signal heads that specifically control bicycle traffic, are mentioned in the *FHWA Study Tour for Pedestrian and Bicyclist Safety in England, Germany, and The Netherlands* (Zegeer et al., 1994).

Chapter 4
Conclusions

The debate over whether BLs or WCLs are preferable has been heated for many years. While both BLs and WCLs are acceptable facilities in many locations, the debate has sometimes forced decision makers to choose which facility type they prefer, to the exclusion of the other. More bicycle facilities might be in place in this country except for this long-standing division of opinion.

This comparative analysis of BL versus WCL sites utilized an extensive data base to examine many factors related to the operations and safety of these facilities. Forty-eight sites from three cities were videotaped in the study, and these produced 369 total conflicts, 276 of which were bike/motor vehicle conflicts. In reality this is relatively few conflicts, which is an encouraging outcome. On the other hand, approximately 6 percent of the bicyclists had a conflict with a motor vehicle, which is not a trivial amount. Many more sites would have been necessary to produce a wholesale increase in the number of conflicts available for analysis.

Across the board these facilities work well, with the vast majority of identified conflicts in this study being minor in nature. Both behavioral actions and geometric characteristics were identified as problems in the study of these bicycle facilities, and there are remedies for these. However, in most cases the noted problems at the higher conflict rate sites could not be labeled as particular BL or WCL deficiencies. The destination patterns of bicyclists traveling through the project sites led to maneuvers and conflicts that in many cases would have occurred whether the bicycle facility present was *either* a BL or WCL.

This is an important point that planners and engineers should heed. With their relative freedom of movement (i.e., not being as confined to a traffic lane as a motor vehicle), bicyclists will use a variety of ways to get through an intersection and on toward their destination. The chosen methods usually reflect perceived time savings/efficiency or improved safety. As an example, difficulties in making left turns because of heavy motor vehicle flows will likely lead to advance crossovers or other alternate maneuvers. Even though standard design templates for bicycle facilities should be applied wherever possible to promote consistency in understanding and proper movements through intersections, it is apparent that such templates cannot be used across the board to achieve standard or desired bicyclist movements. Some tailoring will be necessary to take into account desired or frequent movements by bicyclists, just as is done for locations with high motor vehicle movements and/or crash rates.

The overall conclusion of this research is that *both* BL and WCL facilities can and should be used to improve riding conditions for bicyclists, and this should be viewed as a positive finding for the bicycling community. The identified differences in operations and conflicts were related to the specific destination patterns of bicyclists riding through the intersection areas studied. Given the stated preferences of bicyclists for BLs in prior surveys (e.g., Rodale Press, 1992) along with increased comfort level on BLs found in developing the Bicycle Compatibility Index (Harkey et al., 1998), use of this facility is recommended where there is adequate width, in that BLs are more likely to increase the amount of bicycling than WCLs. Increased bicycling is important because in the United States there are but a few communities that have a significant share of trips made by this mode. Overall, we have not yet reached the critical mass necessary to make motorists and pedestrians aware of the regular presence of the bicycle. When this critical level of bicycling is reached, gains in a "share the road" mentality will come much more quickly than at present. Certainly not all the problems will disappear, but the ability to develop and implement solutions will be greatly enhanced.

References

Harkey, D.L., Reinfurt, D.W., Knuiman, M., Stewart, J.R., and Sorton, A. *Development of the Bicycle Compatibility Index: A Level of Service Approach*, Report No. FHWA-RD-98-072, Federal Highway Administration, McLean, Virginia, 1998.

Harkey, D.L. and Stewart, J.R. "Evaluation of Shared-Use Facilities for Bicycles and Motor Vehicles," *Transportation Research Record 1578*, 1997, pp. 111-118.

Hunter, W.W., Stewart, J.R., Stutts, J.C., Huang, H.H., and Pein, W.E. *A Comparative Analysis of Bicycle Lanes Versus Wide Curb Lanes. Final Report*, Publication No. FHWA-RD-98-034, Federal Highway Administra- tion, McLean, Virginia, 1999.

Hunter, W.W., Stutts, J.C., Pein, W.E., and Cox, C.L. *Pedestrian and Bicycle Crash Types of the Early 1990's*, Publication No. FHWA-RD-95-163, Federal Highway Administration, McLean, Virginia, 1996.

Jensen, S.U., Andersen, K.V., and Nielsen, E.D. "Junctions and Cyclists." *Proceedings of Velo City '97 - 10th International Bicycle Planning Conference Proceedings*, Barcelona, Spain, 1997.

Leden, L. "Has the City of Gothenburg Found the Concept to Encourage Bicycling by Improving Safety for Bicyclists?" *Proceedings of Velo City '97 - 10th International Bicycle Planning Conference*, Barcelona, Spain, 1997.

Leden, L., Gårder, P., and Pulkkinen, U. "Measuring the Safety Effect of Raised Bicycle Crossings Using a New Research Methodology," *Transportation Research Record 1636*, Washington, DC, 1998.

National Bicycling and Walking Study, Publication No. FHWA-PD-94-023, Federal Highway Administration, Washington, DC, 1994.

Pronovost, J. and Lusginan, M. "Behavior of Road Users Following the Application of a Special Bikeway Crossing Marking," *Pro Bike / Pro Walk 96 Resource Book*. Bicycle Federation of America and Pedestrian Federation of America, Portland, Maine, 1996.

Rodale Press Inc., *Pathways for People*, Emmaus, Pennsylvania, 1992.

Zegeer, C.V., Cynecki, M., Fegan, J., Gilleran, B., Lagerwey, P., Tan, C., and Works, B. *FHWA Study Tour for Pedestrian and Bicyclist Safety in England, Germany, and the Netherlands*, Report No. FHWA-PL-95-006, Federal Highway Administration, Washington, DC, 1994.

www.ingramcontent.com/pod-product-compliance
Lightning Source LLC
Chambersburg PA
CBHW081414170526
45166CB00010B/3344